PowerKids Readers:

EARTH MOVERS™

Cranes

Joanne Randolph

New Hanover County Public Library
201 Chestnut Street
Wilmington, NC 28401

For Ryan, with love

Published in 2002 by The Rosen Publishing Group, Inc.
29 East 21st Street, New York, NY 10010

Copyright © 2002 by The Rosen Publishing Group, Inc.

All rights reserved. No part of this book may be reproduced in any form without permission in writing from the publisher, except by a reviewer.

First Edition

Book Design: Michael Donnellan

Photo Credits: p. 5 © CORBIS/Paul Thompson; pp. 7, 19 © SuperStock; p. 9 © DigitalVision; p. 11 © CORBIS/Premium Stock; p. 13 © CORBIS/James Marshall; p. 15 © Highway Images/Bette S. Garber; p. 17 © Highway Images/Genat; p. 21 © CORBIS/Michael S. Yamashita.

Randolph, Joanne.
Cranes / Joanne Randolph.
   p. cm.—(Earth movers)
Includes bibliographical references and index.
ISBN 0-8239-6030-7
1. Cranes, derricks, etc.—Juvenile literature. 2. Cranes, derricks,etc. I. Title.
TJ1363 .R24 2002
621.8'73-dc21

00-013010

Manufactured in the United States of America

Contents

1 What is a Crane?   4
2 What a Crane Does   8
3 Words to Know   22
4 Books and Web Sites   23
5 Index   24
6 Word Count   24
7 Note   24

This is a crane.

Cranes lift heavy things.

This crane helps build a tall building. It lifts heavy things from the ground to the top of the building.

Sometimes it takes many cranes to do a job.

These cranes load and unload ships.

This truck crane is smaller than ones that help build buildings. It can still lift heavy things, though. Look at how it has lifted this big truck off the ground!

A person drives the crane and tells it what to do. Look at all of these levers that the driver uses.

17

This crane has a big claw attached to it. The claw grabs material and moves it somewhere else.

Cranes can do a lot of work.

Words to Know

claw

crane

levers

truck crane

Here are more books to read about cranes:
Construction Trucks (All Aboard Books)
By Jennifer Dussling
Grosset & Dunlap

Diggers and Other Construction Machines (Cutaway Series)
By Jon Richards
Copper Beach Books

Cranes in Action (Enthusiast Color Series)
By Larry Shapiro
Motorbooks International

To learn more about cranes, check out this Web site:
www.howstuffworks.com

Index

B
building(s), 8, 14

C
claw, 18

D
driver, 16

L
levers, 16
lift(s), 6, 8, 14

S
ships, 12

T
truck, 14
truck crane, 14

Word Count: 119
Note to Librarians, Teachers, and Parents

PowerKids Readers are specially designed to help emergent and beginning readers build their skills in reading for information. Simple vocabulary and concepts are paired with photographs of real kids in real-life situations or stunning, detailed images from the natural world around them. Readers will respond to written language by linking meaning with their own everyday experiences and observations. Sentences are short and simple, employing a basic vocabulary of sight words, as well as new words that describe objects or processes that take place in the natural world. Large type, clean design, and photographs corresponding directly to the text all help children to decipher meaning. Features such as a contents page, picture glossary, and index help children get the most out of PowerKids Readers. They also introduce children to the basic elements of a book, which they will encounter in their future reading experiences. Lists of related books and Web sites encourage kids to explore other sources and to continue the process of learning.

Re-added
4-09

5/03

mL